TEMPERATE CLIMATES

Keith Lye

THE WORLD'S CLIMATES

COLD CLIMATES
DRY CLIMATES
EQUATORIAL CLIMATES
TEMPERATE CLIMATES

Produced for Wayland (Publishers) Limited
by *specialist* publishing services, London

Series designer: Eric Drewery
All illustrations Timothy Lole except p39, Alistair Wilson

First published in 1996 by Wayland (Publishers) Limited,
61 Western Road, Hove
East Sussex BN3 1JD England

Cataloguing in Publication Data
Lye, Keith, 1933
   Temperate Climates. - (The World's Climates) 1
Climatic zones - Juvenile literature 2 Mediterranean-type
ecosystems - Juvenile literature
I Title
574.5'264

ISBN 0 7502 1819 3

Printed and bound by G. Canale, Turin, Italy

Cover picture: Sourwood tree along Unicoi Lake, Unicoi
State Park, Georgia, USA, Tony Stone Images.

Picture acknowledgements
Carol Kane 19, 32–33, 36; FLPA 16, 28;
Robert Harding Picture Library 1, 5, 9, 11,
13, 15, 16–17, 19, 20, 21, 22, 23, 25, 26, 27,
31, 32–33, 34–35, 37, 38, 39, 40, 41, 42, 43;
Trip 5, 7, 11, 17, 27, 31, 35, 38–39.

Contents

The middle latitudes

CLIMATIC REGIONS

Climate is the usual, or average, weather of a place. It determines what kinds of plants and animals are found in an areas, and it influences how people live. The two main factors that determine climate are temperature and precipitation. Precipitation includes rain, snow, sleet, hail, dew and frost – in fact all forms of moisture that come from the air.

Places with temperate climates occur in the middle latitudes – between the low latitudes of the hot tropics and the high latitudes of the cold poles. They have four distinct seasons, so they do not suffer the extreme cold of polar and subarctic climates, nor do they have the constant, year-round heat of tropical regions. The precipitation is generally plentiful, so temperate regions do not include dry climates, even though

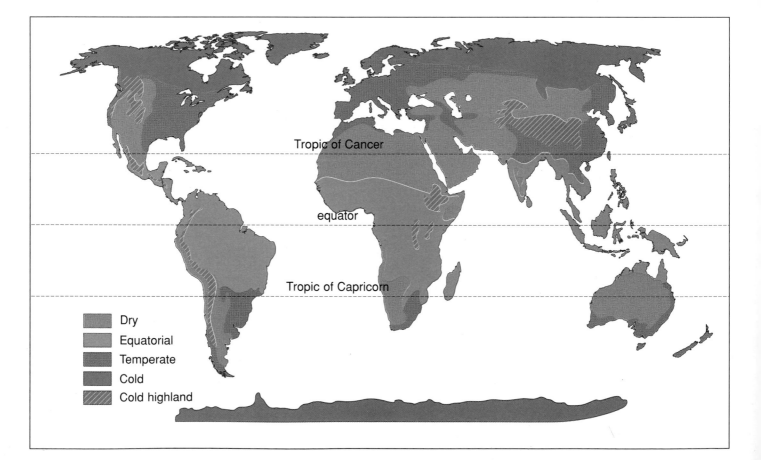

Tropic of Cancer

equator

Tropic of Capricorn

Dry
Equatorial
Temperate
Cold
Cold highland

Above Sunbathers on Bondi Beach, Sydney, Australia.

Right Red Square, Moscow, in the winter.

Left The main climate regions of the world.

some deserts and dry grasslands do occur in the middle latitudes.

There are three main types of temperate climates. The first type includes places with hot, dry summers and mild, wet winters. They are said to have a Mediterranean climate, after the kind of climate that occurs around the Mediterranean Sea. Mediterranean climates also occur in smaller areas in North and South America, Australia and South Africa.

The second type of temperate climate is the rainy temperate climate. It has rain throughout the year. This climate occurs in the south-western United States, much of eastern China and southern Brazil.

The third type of temperate climate is the cold temperate climate. Places with this climate have cold winters and hot summers. They occur in the hearts of continents, far from the moderating effect of the sea. For example, the town of Stirling in Scotland lies on roughly the same latitude as Moscow in Russia. Yet the average temperature in Stirling in January is 3 °C, while in Moscow it is −8 °C.

FOUR SEASONS

The seasons in temperate regions are caused by the tilt in the earth's axis. The axis is the imaginary line through the earth that joins the North Pole, the centre of the earth and the South Pole. As the earth travels around the sun, areas of the earth tilt towards the sun, getting more sunlight, while other parts tilt away, getting less sunlight.

On March 20 or 21, the sun is directly overhead at the equator (the imaginary line that runs around the earth, half way between the two poles, which divides the world into the northern and southern hemispheres).

March 20 or 21 is the spring equinox in the northern hemisphere. At the spring equinox, night and day are roughly 12 hours long all over the world.

After March 21, the northern hemisphere tilts increasingly towards the sun, until, on June 20 or 21, the sun is overhead at the Tropic of Cancer (latitude $23\frac{1}{2}°$ north). This is the summer solstice in the northern hemisphere.

After June 21, the northern hemisphere starts to tilt away from the sun.

By September 22 or 23, the sun is again overhead at the equator. This is the autumnal equinox in the northern hemisphere. At the autumn equinox, night and day are again roughly 12 hours long all over the world.

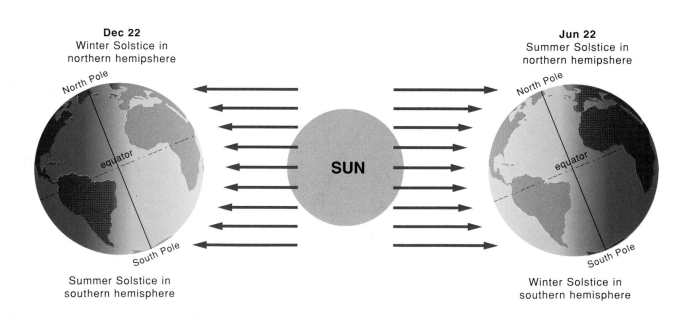

Dec 22
Winter Solstice in northern hemipshere

North Pole

equator

South Pole

Summer Solstice in southern hemisphere

SUN

Jun 22
Summer Solstice in northern hemipshere

North Pole

equator

South Pole

Winter Solstice in southern hemisphere

How the sun affects the summer and winter seasons.

After September 23, the northern hemisphere starts to tilt away from the sun.

Finally, on December 20 or 21, the sun is overhead at the Tropic of Capricorn (latitude $23\frac{1}{2}°$ south).

This is the winter solstice in the northern hemisphere.

But December 20 or 21 is the summer solstice in the southern hemisphere. The seasons are reversed south of the equator.

The same scene in summer (left) and winter (right) – old packhorse bridge at Carrbridge, Scotland.

DEPRESSIONS AND ANTICYCLONES

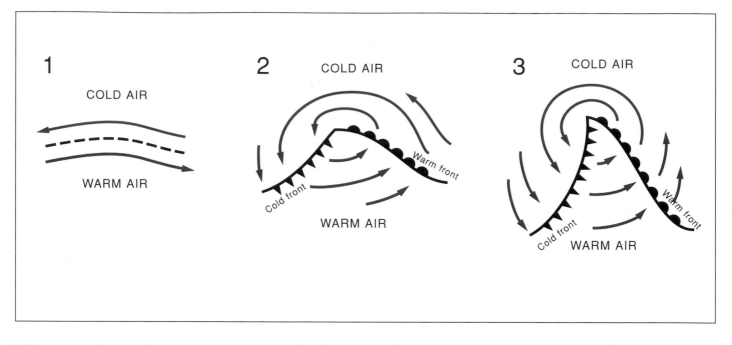

The weather changes a great deal in places that have a temperate climate, such as Britain. The reason for the changes is that the climate is influenced by two weather systems: anticyclones and depressions (or cyclones).

Anticyclones are regions of high air pressure where air is sinking. They bring stable weather. In winter, anticyclones bring long spells of cold, frosty weather. In summer, they bring blue skies and sunshine. By contrast, depressions are regions of low air pressure where warm air rises, clouds form in the rising air and rain or snow are common.

Warm westerly winds, blowing from subtropical zones, meet up

Above Depressions form along a polar front.

Right A sunny day and a cloudless sky – the result of an anticyclone – Brighton, England.

Below This map of an anticyclone shows how the highest air pressures are at the centre, with the winds blowing outwards.

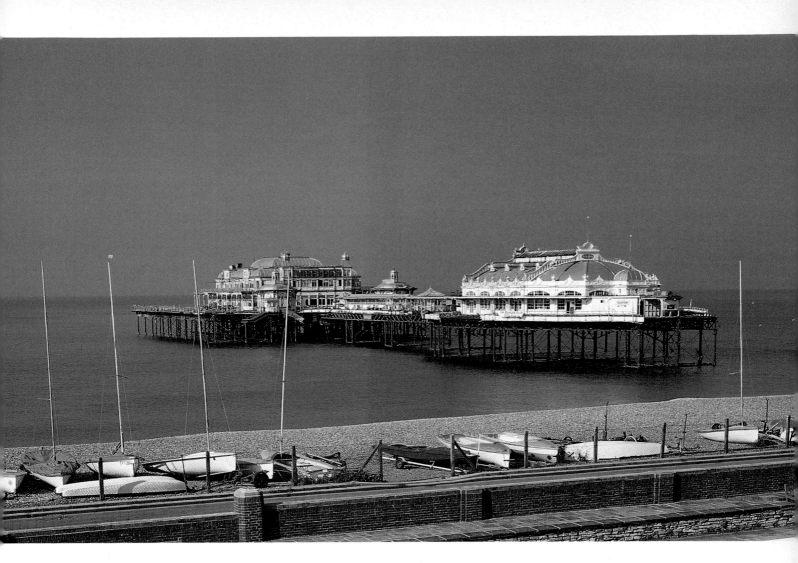

with cold easterly winds blowing from polar regions at the 'polar front'. This is how depressions form. Because warm and cold air do not mix easily, the warm air flows into bends in the polar front. Cold air flows in behind the warm air. This sets up a circular air system or depression. At the centre of the depression is the lowest air pressure.

So depressions are marked by fronts. The warm front is the edge of the advancing warm air and the cold front is the edge of the cold air flowing in behind.

Along both fronts are bands of cloud, which bring stormy weather. The rain that falls during depressions is called cyclonic rain.

Eventually the cold, heavy air overtakes the warm, light air. The places where this happens are called occlusions. Along occlusions, cloud and rain persist for some time.

When, eventually, the warm air becomes cold, the depression dies out.

CHANGEABLE WEATHER

As depressions move across the land, they bring changing weather conditions. Depressions vary greatly in size, measuring anything between 150–3000 km across.

They often move rapidly from west to east across the temperate regions in the northern hemisphere. Many move at around 1000 km per day, but some are slow-moving or almost stationary.

Warm air contains invisible water vapour. When warm air rises, it cools. Cold air cannot hold as much water vapour as warm air can. As a result, the water vapour condenses, turning into tiny crystals of ice or droplets of water. Billions of ice crystals and/or water droplets form clouds.

The first sign of an approaching depression may be wispy, high clouds (cirrus), which form as the warm air rises above the cold air. Cirrus clouds are made up of tiny ice crystals.

Other clouds follow. Just ahead of the warm front are dark grey, flattish clouds (nimbostratus). These bring steady rain or snow.

When the warm front arrives, the rain eases. Temperatures rise and the air becomes more humid.

However, after a few hours (or a few days if the depression is moving slowly) the cold front arrives.

Along the cold front, which moves faster than the warm front, the cold air undercuts the warm. This pushes the warm air upwards and high clouds form in the rising air, creating thunderclouds (cumulonimbus). Thunder and lightning, heavy rain, hail, or sleet and snow in winter are features of thunderclouds.

These storms do not normally last for long and the cold front soon passes away. The skies clear and temperatures fall.

Above *This weather station supplies data which is used to produce weather forecasts.*

Left *A diagram of a depression, which is moving to the right as you see it.*

Right *Heavy rain has caused the flooding of this rugby field.*

OCEAN CURRENTS AND CLIMATE

Ocean currents are the movements of water in the oceans. Onshore winds that blow across the currents are either warmed or chilled, depending on whether the current is warm or cool. This affects the climate of land near the sea.

Warm currents that flow from the tropics towards the polar regions warm them up. Cold currents that flow from polar regions to the tropics cool them.

The Gulf Stream, a current that starts in the Gulf of Mexico, has a great influence on European climates. When this warm-water current emerges through the Strait of Florida, it flows up the east coast of the United States and then across the Atlantic.

Above *Diagram showing the warm and cool currents of the North Atlantic Ocean.*

Part of the current turns back down the coast of Africa as the Canary Current.

Another part forms a slower current called the North Atlantic Drift. This affects the climate of

The warming effect of the Gulf Stream means that the winters in Brest, France are mild, as this temperature graph shows.

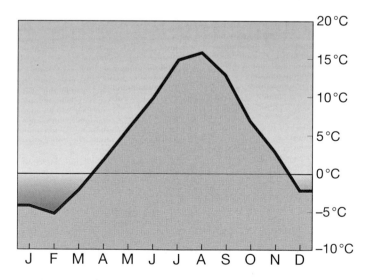

St John's, Newfoundland, further south than Brest, is not warmed by ocean currents as Brest is, and so has harsh winters.

Above *The port of Narvik, Norway, which remains ice-free throughout the year because of the warm ocean current that flows nearby.*

such places as Iceland, the British Isles and Scandinavia. It keeps most of Norway's seaports ice-free throughout the year – even the port of Hammerfest, 460 km north of the Arctic Circle, is always ice-free.

The influence of the Gulf Stream and the North Atlantic Drift also explains why the climate of north-western Europe is much milder than the climate of eastern Canada.

For example, Brest in France is farther north than St John's in Canada, but Brest has much warmer winters than St John's. The climates of the east and west coasts of continents are often very different. This is because warm currents usually flow along the east coasts of continents, while cold currents normally flow along the west coasts.

Mediterranean climates

DRY SUMMERS, WET WINTERS

For summer holidays, many people want to find a seaside resort where they can be sure that the sun will shine for most of the time. Many popular tourist resorts are in places with Mediterranean climates. This is because summers are sunny and dry, with average temperatures rising to 21 °C or, in some cases, as high as 28 °C. The mild winters are another attraction, especially to retired people who were brought up in countries with snowy winters. In Mediterranean regions, the average temperature in the coldest month is between 6–10 °C.

Mediterranean climates occur in most of the lands around the Mediterranean Sea, though the coasts of Egypt and Libya have desert climates. Mediterranean climates also occur in smaller areas in central California, central Chile, the south-western tip of South Africa and parts of southern Australia. These regions have much in common. They are all known for their orange and lemon groves, vineyards, wine-making industries and tourist beaches. In most areas, the original forests have been largely cut down and replaced by low scrub or open heath. Fire is a constant danger, especially during the dry summers.

Regions with Mediterranean climates have an average annual rainfall of between 400–900 mm. Most of the rain comes in winter

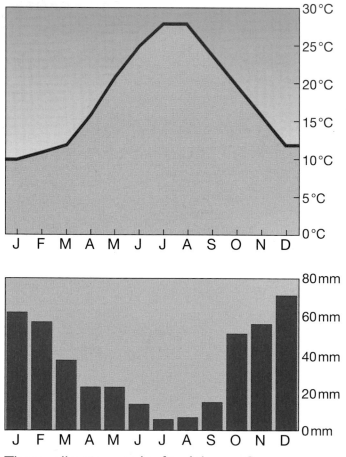

These climate graphs for Athens, Greece show its dry, warm summers and mild winters.

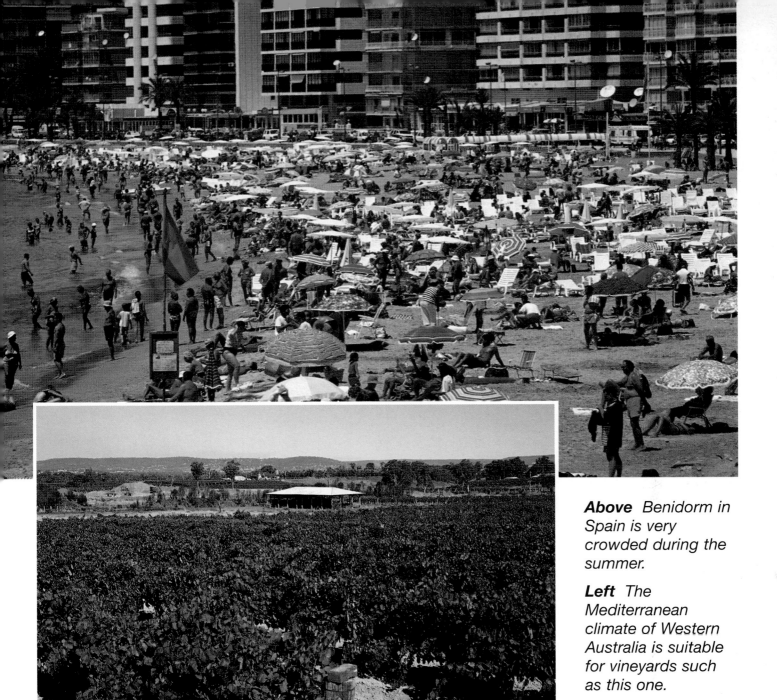

Above Benidorm in Spain is very crowded during the summer.

Left The Mediterranean climate of Western Australia is suitable for vineyards such as this one.

from depressions that move into the area.

In summer, Mediterranean climate regions come under the influence of high air-pressure systems. When the land is intensely heated and hot air rises in strong currents, water vapour in the rising air condenses (turns into droplets of water) and forms thunderclouds that bring heavy rain.

However, storms do not last long and afterwards it is usually pleasantly cool.

PLANTS AND ANIMALS

Forests of mainly evergreen trees once grew over most of the land in Mediterranean climate regions. Mild winters enable plants to grow throughout the year, without a winter pause in growth. But, from ancient times, the original forests were cut down. Many people around the Mediterranean Sea were sailors who needed timber to build boats. Other people cleared forests to create farmland. In many areas, goats and other animals over-grazed the land, killing off the natural vegetation. Over large areas, fire has also restricted plant growth.

Some woodlands of chestnut and evergreen oak, including the valuable cork oak, still survive

Left Olive trees are evergreens – this one is in Israel.

Below Wild horses live in the Camargue nature reserve in southern France.

Below far right Flamingoes are also part of the wide variety of life in the Camargue.

around the Mediterranean Sea, but scrub or heath is now the main vegetation. The original forests in other regions with Mediterranean climates have also largely disappeared and replaced by scrub. The scrub has many names. In France it is called the maquis or garigue, in Italy the macchie, in California the chaparral, in Australia the mallee and in South Africa the fynbo. The scrub is dominated by shrubs, such as the proteas of South Africa and Australia, the sagebrush of California and the myrtles and rockroses of Europe.

Animal life is rich. For example, roe deer are widespread in parts of Mediterranean Europe. However, many other animals, such as lynxes and wolves, are now rare. Jackrabbits and pumas live in California's chaparral. Kangaroos are the commonest grazers in Australia. Duikers and hyraxes live in the South African fynbo.

The Mediterranean Sea region has many snakes and other reptiles. There are a wide variety of lizards. Birds include many species of warblers. The region is also the winter home of many birds that breed in the colder parts of northern Europe and then migrate. Smaller mammals include such creatures as dormice, foxes, hedgehogs and rabbits.

THE MEDITERRANEAN SEA REGION

Mediterranean climates affect about 1.7 per cent of the world's land area. The largest single area is in the Mediterranean Sea basin, a region that has been occupied by people since ancient times. It gave birth to some great civilizations, including the Minoan, ancient Greek and Roman. Much of the land has been greatly affected by human activities.

Apart from the interior uplands and the dry coasts of north-eastern Africa, the lands around the Mediterranean Sea typically have hot, dry summers and mild, wet winters.

The effect of the winter depressions, which come from the Atlantic Ocean, is greater in the western countries. Generally speaking, rainfall decreases from west to east throughout the Mediterranean Sea.

The wetter areas have some woodland, though most trees are low or stunted. The typical vegetation consists of a dense cover of low bushes, including laurel and myrtle. Drier areas have sparser vegetation. Here grasses, with their shallow roots, struggle to survive during the hot summers. Common plants include the sweet-smelling broom, lavender, rockrose, rosemary and thyme.

In the winter, cold winds blow in winter from the interior of Europe. For example, the bora brings cold weather to northern Italy and the east coast of the Adriatic Sea.

Climate graphs show Gibraltar's mild winter.

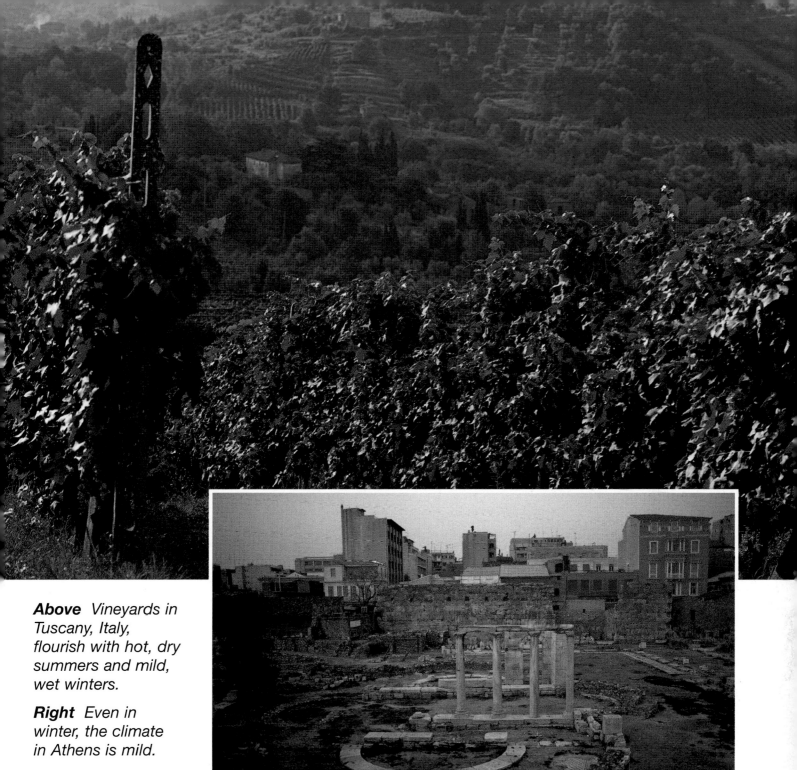

Above Vineyards in Tuscany, Italy, flourish with hot, dry summers and mild, wet winters.

Right Even in winter, the climate in Athens is mild.

In France, the mistral is a strong wind that blows south down the Rhône valley, bringing cold and dry weather to the Rhône delta. Hot winds blow from the Sahara.

The sirocco is a dry, dusty wind at the Algerian coast, but it picks up moisture over the sea and when it reaches Sicily and southern Italy it is hot and unpleasantly humid.

THE AMERICAS

The central valley and southern coastal regions of California, and central Chilein South America, both have Mediterranean climates.

However, the coastal regions in both California and Chile have much cooler summers than inland areas. They are cooled by onshore winds that blow over cold ocean currents. These cool winds also bring sea fogs.

The inland valleys, that run north–south behind coastal ranges, are much warmer. For example, Sacramento, in the central valley of California, has an average temperature of 23 °C in July. This is compared with 15 °C in San Francisco which is on the coast.

The cool onshore winds do sometimes affect the central valley when they are drawn inland through gaps in the coastal mountains.

In these regions, winters are mild. Sacramento has an average January temperature of 8 °C.

However, nights are often cool. Frosts occasionally occur and fires are sometimes lit in Californian orange groves to prevent frost damage. The rainfall is moderate, most of it coming from depressions that cross the regions in winter.

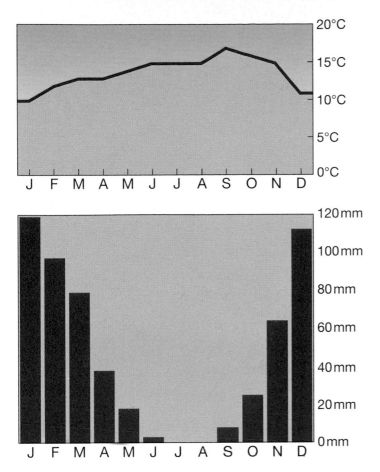

Both central California and central Chile are important farming areas which produce such crops as citrus fruits and grapes.

There are also areas of scrub, with wild lilac, highland live oak and sumacs (plants of the cashew family) being common plants in the Californian chaparral. Redwood trees and yellow pines grow on the wetter northern coast-ranges of California, while forests of Araucaria pine grow on the slopes of the Andes. In parts of both California and central Chile, the Mediterranean regions merge into areas of steppe or desert with dry climates.

Above *Climate graphs for San Francisco show mild coastal summers that are caused by sea fog.*

Left *The Californian coast is one of the foggiest parts of the United States. Here The Golden Gate Bridge, San Francisco, is shrouded in fog.*

Right *A farmer tends his crops in Chile, South America.*

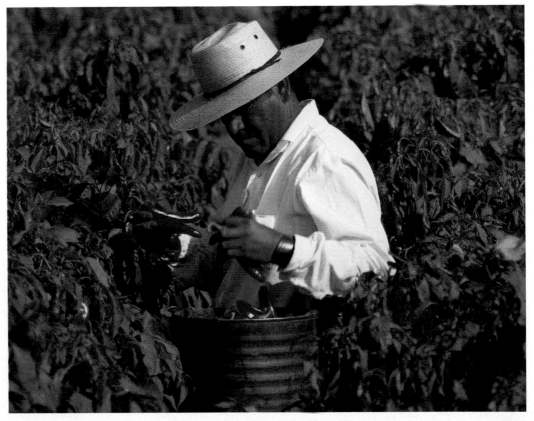

SOUTH AFRICA AND AUSTRALIA

Mediterranean climates occur in three other areas. All of them are well-known for producing Mediterranean crops such as citrus fruits, wheat, grapes and quality wines.

These regions include the southern tip of South Africa, around the city of Cape Town. Most rain comes in winter from depressions, though occasional thunderstorms bring rain in summer. The area has warm summers, with average temperatures reaching 21 °C in Cape Town in January. Winters are mild and the average temperature in July (the coldest month) is 12 °C.

Occasionally a warm wind, called the berg wind, reaches the area from the interior. It can raise temperatures to 38 °C.

Two other regions with Mediterranean climates are in southern Australia. They include the south-western tip of Australia, around Perth, while a second area includes the Adelaide district and part of Victoria. Between these two regions is a dry area bordering the Great Australian Bight. The winter depressions that bring rain to Australia's Mediterranean climate areas pass south of the Bight.

South-western Australia has a mild, almost frost-free climate. The average

Right Protea, the national emblem of South Africa, grows in the Mediterranean climate of the Cape of Good Hope.

Right A sunny day in Perth, Western Australia. During winter depressions bring rain to Perth, making it even wetter than Adelaide.

Climate graphs for Adelaide show rainfall throughout the year, with most in winter.

temperature in Perth in the coldest month (June) is 13°C. Perth has an average annual rainfall of about 880 mm, most of which is brought by depressions in winter, between May and August. Adelaide is drier. Some rain falls in summer but the winter months are still the wettest.

The plant life in these regions changes from the wetter coastal areas to the drier inland areas.

In south-western South Africa, the remains of coniferous forests still grow but fynbo scrub is the main kind of natural vegetation.

In Australia, the wetter coastal areas contain some forest but mallee scrub is dominant inland. Mallee is the name for a dwarf, many-stemmed eucalyptus tree.

Rainy temperate climates

WET AND WARM

There is another group of temperate climatic regions that has rain throughout the year. This group is divided into three main types, classified according to their temperatures in summer. One group has hot summers, another has warm summers and a third has cool summers.

Temperate rainy climates with hot summers occur mostly on the eastern sides of continents, where the coastal areas are influenced by warm ocean currents. These regions are often described as subtropical even though they lie outside the tropics, because they have mild or warm winters. Rain occurs in every month, but summer is often the wettest season. Much of the rainfall occurs during thunderstorms, when intense heating of the land causes warm air to rise and then drop its moisture. Hurricanes also bring rain in summer, but in winter rain is often caused by depressions. The south-east coast of Queensland, Australia, around the city of Brisbane, has a climate of this type. Other areas include the south-eastern United States, much of China, southern Japan, part of South America (including southern Brazil) and South Africa's south-east coast.

Brisbane's graphs show wet, warm summers.

Right Berlin, Germany has a rainy temperate climate with warm summers.

Below Melbourne is in south-east Australia, which has a rainy, temperate climate with warm summers.

Regions with temperate rainy climates and warm summers occur in much of western Europe, including the British Isles. Here, the variable weather is caused by the influence of depressions and anticyclones, which occur throughout the year. Other temperate rainy regions with warm summers include south-eastern Australia and western Oregon and Washington in the north-western United States.

The third group of temperate rainy climates, with cool summers, includes western British Columbia in Canada, south-western Norway, where temperatures are raised by the warm North Atlantic Drift and southern Chile. Christchurch in New Zealand also has a climate of this kind.

RAINY WITH HOT SUMMERS

Temperate climates with hot summers and no dry season are ideal for the growth of forests. Winters are mild or even warm and snow is unusual.

Trees continue to grow throughout the year so evergreens, such as cypresses, laurels and evergreen oaks, are among the many trees found in these regions. Despite the cutting down of trees to create farmland, forests are still found in these regions.

In parts of the south-eastern United States, there are mixed evergreen and pine forests, with pine forests along the sandy coastal plains.

The south-eastern United States contains several swamp regions, one of which is the Everglades in Florida. This famous region has many thickly wooded islands covered by dwarf cypress and pine, and mangrove forests near the coast. It supports three hundred kinds of birds, and six hundred kinds of mammal, including deer, otters, panthers and raccoons. The king of the swamp is a reptile – the American alligator.

Southern Brazil, northern Argentina and Uruguay also have a rainy climate with a hot summer.

Left The hurricanes which bring rain to Louisiana, United States, also cause damage such as this.

Top right Alligators live in the Florida Everglades, United States.

Bottom right In Suzhou Creek, Shanghai, China, the average temperature in the warmest month is 27°C.

In these regions, forests probably once covered large areas, but they have been cut down or destroyed by fire. As a result, only a few patches of forest survive and the region is mainly grassland.

East-central China has some evergreen forests in the south, and mixed forests of evergreens and deciduous trees farther north. As in South America, large areas have been cleared for farming.

Because of the climate, warm-weather crops grow well in these regions. For example, the south-eastern United States produces cotton, sugar-cane and tobacco.

This region has also become a resort area, especially in winter, when thousands of visitors arrive from the cold north-east.

RAINY WITH WARM SUMMERS

The British Isles and much of west-central Europe make up the largest region with a temperate rainy climate and warm summers. The highest temperatures occur in the summer, when high pressure brings hot air from the south and east (south-westerly winds are warmed by the Gulf Stream). In winter, high pressure brings cold air from the north and east and sometimes causes long spells of freezing weather.

Most rain in Western Europe is from depressions that come from the Atlantic Ocean. The wettest and coolest areas are in the west, especially on highlands.

Climate graphs show Berlin's temperate rainy climate with warm summers.

Left *The marked differences between summer and winter in the British Isles meant that deciduous forests once covered large areas. Deciduous trees (ash, beech, birch, chestnut and oak) shed their leaves in winter. Most of those forests have now been replaced by farm and grazing land.*

RAINY WITH COOL SUMMERS

Above *Christchurch, New Zealand, has a temperate rainy climate with cool summers.*

The third group of temperate rainy climates has cool summers. Regions with this type of climate lie in the high latitudes but they are close to the sea, which moderates their climates. For example, south-western Norway would have a much cooler climate were it not for the onshore winds which are warmed as they pass over the North Atlantic Drift, the northern extension of the Gulf Stream.

The climate of western British Columbia, Canada, is also affected by mild winds that blow onshore, bringing heavy rain. But the moderating effect is not so great as in south-western Norway and, as a result, western British Columbia has colder winters than south-western Norway.

A third temperate region with cool summers is the southern part of Chile, where winds from the oceans bring storms, but they also moderate the temperature throughout the year.

The south-east coasts of South Island, New Zealand have a similar temperate, though less wet, climate.

29

Cold temperate climates

CONTINENTAL MOIST CLIMATES

Some large areas in the middle latitudes have warm or hot summers, but winters that are cold and snowy. These places have a cold temperate climate. During the long winters, it is often as cold as in subarctic regions, and frosts are common in spring and autumn. Because the frosts shorten the growing season, crops must be hardy to survive.

In summer, heatwaves often occur. They are sometimes severe. In July 1995, a heatwave in the north-central and the north-eastern United States sent temperatures soaring above 40 °C. Lasting nearly a week, it caused the deaths of more than 370 people in Chicago. The heatwave also claimed victims in Philadelphia, New York City and Washington D.C. Many of the people who died were elderly and most of them lived in homes with no cooling system.

Most regions with cold temperate climates lie far from the sea, in the hearts of continents. But some coastal areas, where cold ocean currents chill onshore winds, also have cold temperate climates. For example, south-eastern Canada is

Climate graphs for New York City, where cold ocean currents reduce winter temperatures.

cooled by the Labrador Current. The icy water in this current comes from the Arctic, often bringing jagged icebergs into the North Atlantic Ocean.

Other areas with cold temperate climates include the north-eastern and mid-western United States, eastern Europe and west-central Russia, and north-eastern China. The islands of Hokkaido and northern Honshu in Japan also have a cold temperate climate. These islands are chilled by the cold Oya Current, which comes from the Arctic.

No cold temperate regions occur in the southern hemisphere. This is because the higher middle latitudes south of the equator do not contain any large continental areas.

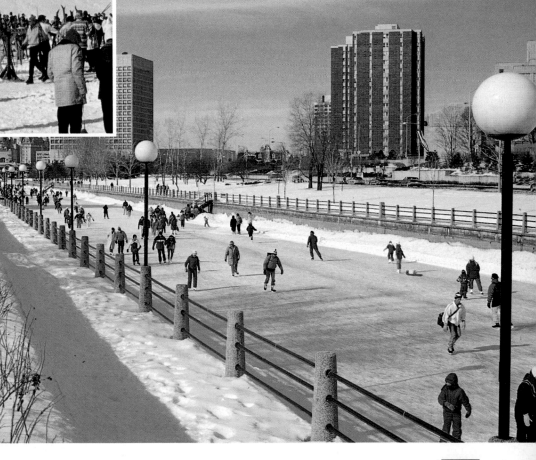

Above *The long, cold winters in Denver, Colorado, make it a popular resort for skiers.*

Right *Skaters on the Rideau Canal, Ottawa. Canada has a more severe climate than north-eastern United States, with a temperature range of −11° C (January) to 21°C (July).*

NORTH AMERICA

North America contains a large cold temperate region. It lies between the subarctic coniferous forest zone in Canada and the warm temperate region in the south-eastern United States. It includes the Great Lakes region and the St Lawrence River valley, New England and New York state. In the Mid-West it merges into the dry prairie regions.

In winter, the cold temperate region comes under the influence of a large high air-pressure system which tends to turn away depressions. This reduces the winter precipitation. Instead, the summer months are the wettest. Rain comes from depressions and also from thunderstorms which

Above The natural display of the autumn vegetation in Conneticut, New England, attracts many visitors.

Left The Great Lakes in Canada never completely freeze over. This is Lake Ontario.

Right Wheat is grown on the dry prairies of the Mid-West.

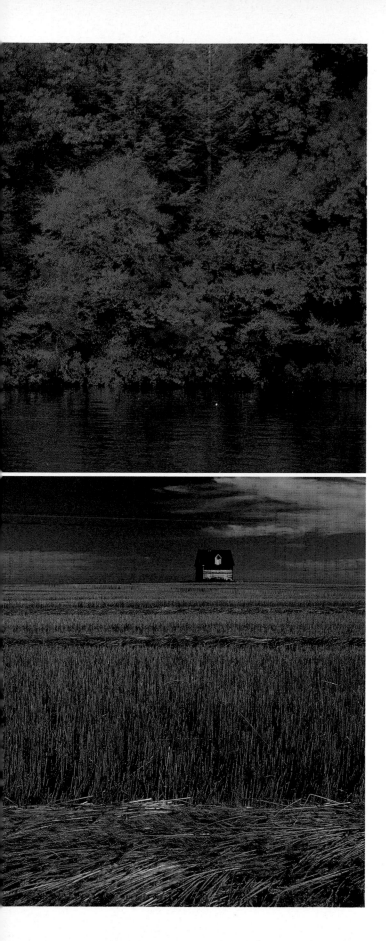

form on hot afternoons. Generally, the rainfall decreases from east to west. Boston, Massachusetts, has an average yearly precipitation of about 1040 mm, while Chicago, to the west, has 840 mm.

The coldest parts of this region are in the north, but almost everywhere average temperatures, are below freezing point for at least one month a year. The Great Lakes help to moderate temperatures, and mild spells of weather often occur even in mid-winter. However, there are also spells of bitterly cold weather when waves of polar air sweep south. These cold winds often bring fierce blizzards.

Unpleasant conditions also occur in summer. Heatwaves occur from time to time, when daytime temperatures may rise to 38 °C for several days.

Deciduous and coniferous trees once grew over much of this area, but most of the original forest has been cut down.

However, the mixed forests that grow in New England provide one of the world's finest spectacles. In autumn, the leaves of the deciduous trees change colour, creating a blaze of yellow, gold, brown, scarlet and pink, against a background of green from the coniferous trees.

WEST-CENTRAL RUSSIA

Another region with a cold temperate climate extends from eastern Europe through west-central Russia into Siberia. In the west, it includes southern Sweden, southern Finland, Estonia, Latvia, Lithuania, Belarus and parts of Poland, Hungary and Romania. In Russia, the region lies between subarctic, coniferous forests to the north and the dry steppes to the south.

Climate graphs for Moscow show severe winters, warm summers and moderate rainfall.

Winters are mildest around the Baltic Sea in the south-west. To the east, winters become increasingly severe. In west-central Russia winters are cold, much like those in southern Canada. For example, Moscow has five months when the average temperatures are at or below freezing. The cold winters have had an impact on Russia's history – they were an important factor in the defeats of Napoleon's French armies in the 1810s and Hitler's armies in the 1940s. In both cases, Russian troops were

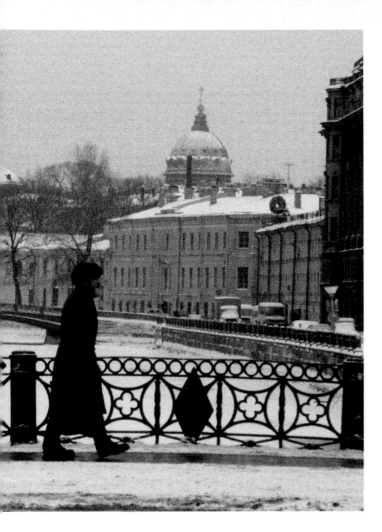

Mixed forests of deciduous and coniferous trees grow in wetter parts of this region, with grasslands in drier areas. The forests once teemed with wildlife, but many animals have been hunted almost to extinction. The Bialowieza Forest on the border between Poland and Belarus gives a good idea of what these forests were like. It contains such trees as alder, ash, hornbeam, lime, oak and pine, and is the home of the rare wisent (European bison).

Left *Winters in St Petersburg, Russia, are harsh and cold.*

Below *Rye, oats and potatoes are crops that can grow during the short growing season of central European Russia. This grain silo for loading lorries is at Krasnodar.*

more able to cope with the harsh winters than their enemies.

Temperatures rise rapidly in spring, and summers are pleasantly warm, becoming warmer to the east and south. The precipitation is moderate, occurring throughout the year, but the largest amounts fall in summer. In winter, the precipitation occurs mainly as snow. Even in Stockholm, in southern Sweden, snow covers the land for about 60 days every year. The summer rainfall comes from depressions and thunderstorms.

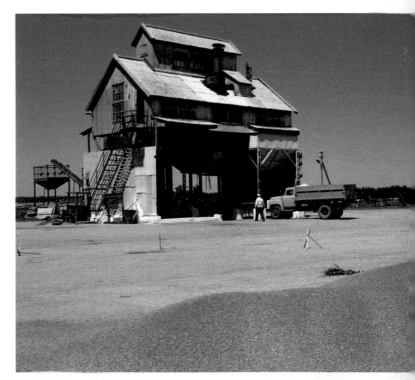

Changing environments

CITY CLIMATES

Every place on earth has its own climate. Climatologists (people who study climate) have divided the world into large climatic regions, including the temperate regions described in this book. However, even tiny areas, such as gardens, have their own climate. For example, a gently sloping garden facing south in the northern hemisphere has a slightly warmer climate than a sloping garden facing north. The study of the climate of small areas is called microclimatology – 'micro' comes from the Greek word meaning 'small'.

Most people in temperate latitudes live in cities and towns. Many live and work in buildings which are centrally heated in winter and air conditioned in

Left Winds are channelled along this 'canyon-like' street in New York.

Right Koshosdale power station in Kentucky, United States, releasing heat into the atmosphere.

Far right A tarmac road melts in the heat. Tarmac absorbs more solar radiation than grasslands or forest. In a city, this absorbed heat from the roads can raise the air temperature.

summer. This means that the insides of the buildings have an artificial climate.

Heating systems, together with cars and other vehicles, factories and power stations, help to make cities warm places.

Also, the walls of buildings, stone pavements and tarmac roads absorb heat from the sun. These surfaces then heat the air around them to a much greater extent than the fields and farms around the cities. Hence, climatologists describe cities as 'heat islands' set in the cooler 'sea' of the surrounding countryside. In winter, for example, cities often have fewer frosts than the surrounding countryside, and people in cities often have to pay less for their heating bills. The differences between temperatures in the cities and the countryside are greatest on working days and least at weekends.

City buildings slow down winds blowing across the land. The moving air is sometimes funnelled in between buildings, which produces strong gusts which can be very unpleasant.

CLIMATE AND POLLUTION

The activities of people in cities cause pollution, which affects climate. City factories and power stations burn fuels such as coal, gas and oil. Smoke from factories and power stations contains a gas called sulphur dioxide. Other gases, called nitrogen oxides, are produced by petrol-run vehicles.

In the air, sulphur dioxide and nitrogen oxides are dissolved by the tiny water droplets that make up clouds. This turns them into sulphuric and nitric acids. When the water droplets merge with others to form raindrops they

Above A coal-fired power station emits sulphur dioxide into the atmosphere.

Left Petrol-run cars emit fumes, including nitrogen oxide, into the atmosphere.

Right Acid rain has killed these forest trees in Bohemia in the Czech Republic.

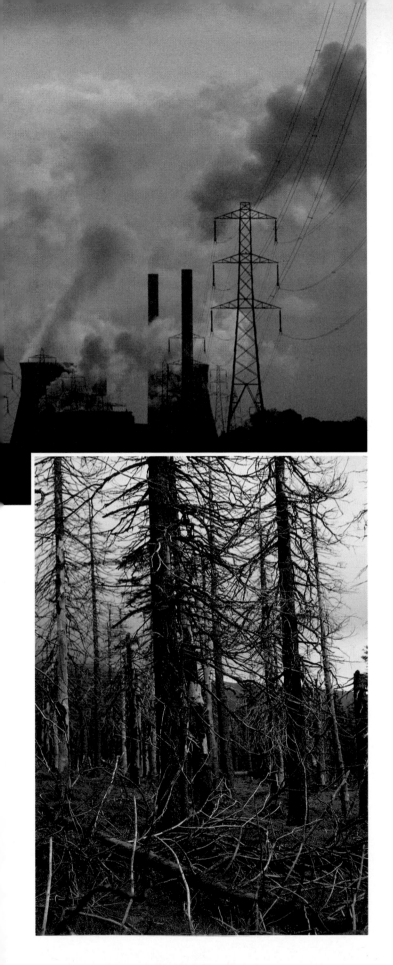

become acid rain. Because the clouds are blown by the wind, the acid rain often falls a long way away from the cities and factories. For example, industries in the north-eastern United States may be responsible for acid rain in south-eastern Canada.

All rain is slightly acidic because it has dissolved some carbon dioxide from the air, but acid rain is much stronger and highly destructive when it reaches the ground. When it enters lakes and streams, it kills off fish and other living things. It also wears away stonework, such as the delicate stone carvings that decorate churches.

The most obvious effect of acid rain is the damage it has done to trees. Acid rain has destroyed large areas of forest in eastern North America, in north-western and central Europe, and in parts of Asia.

Some countries have passed laws to try to reduce the amounts of sulphur dioxide and nitrogen oxides released into the air, which is the only way to reduce the amount of acid rain.

In some areas, lime is added to lakes and streams. This stops the harmful effects of the acid, but is only a temporary measure and it may have harmful side effects.

VANISHING FORESTS

In Germany, people say that trees attacked by acid rain are suffering from *Waldsterben* ('forest death'). But acid rain has only become an important factor in killing off trees in the last 50 years. Most of the original forests of the temperate regions were destroyed long ago.

Above *Bialowieza National Park, Poland.*

Above right *A Cedar of Lebanon – only a few of these trees have survived, because their timber was used to build ships.*

Above far right *A 'clean air' demonstration against acid rain.*

Left *Great Smoky Mountains National Park, North Carolina, United States.*

In Europe, in ancient times, farmers cleared the land to grow crops. They also used trees to build houses and ships. The destruction of forests was so widespread that very little original forest survived. For example, the New Forest was preserved because it became a hunting place for Norman kings.

In what is now the north-eastern United States, many Native American groups, such as the Chippewa, Huron and Iroquois, lived in the mixed forests. They hunted, fished and grew crops in forest clearings. When the soil in a clearing became infertile, the people moved on, and soon trees were again growing on the old clearing. But from the early seventeenth century, European settlers began to develop the land. They tore down the forests, cutting down trees for building and to create farmland.

The rich wildlife of the area dwindled because its forest habitats had disappeared.

The settlers exposed the land to the rain and wind. The rain flowed down bare hillsides, wearing out deep gullies and removing the soil. This soil erosion made large areas as barren as the surface of the moon. The soil was washed into rivers, which became choked with mud. The swollen rivers burst their banks, causing floods.

It is only in the last 70 years that large-scale efforts have been made to control soil erosion and conserve forests.

OUR CHANGING WORLD

Around 200 years ago, the world had about 1 million people. Most of them lived in country areas and worked on the land. In the nineteenth century, more and more people moved into towns and cities to work in factories. This change, called the Industrial Revolution, was accompanied by a rapid increase in the world's population. By the mid-1990s, the world's population had reached 5.6 million and the majority of people in temperate regions lived in cities and towns.

The factories and machines of the Industrial Revolution burned coal, oil and natural gas. These are called fossil fuels because they are formed from the remains of once-living things.

When fossil fuels are burned they give off various gases. One of these gases is carbon dioxide, a 'greenhouse' gas. It is called this because it traps heat that is

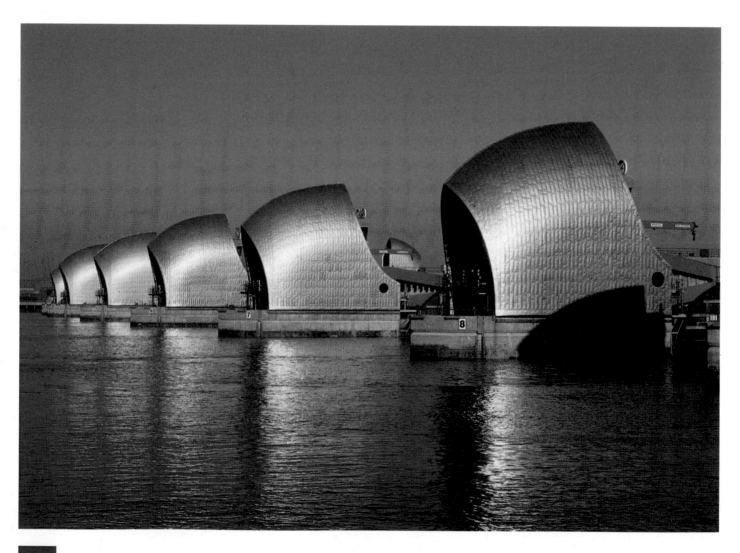

reflected back from the earth's surface, much like the glass in a greenhouse. The more carbon dioxide there is in the atmosphere, the more heat is trapped and the hotter it becomes.

Before 1700, carbon dioxide made up about 280 parts per million in the atmosphere. Nowadays it has risen to 355 parts and is still rising. As a result, world temperatures have risen by an average of between 0.3–0.6 °C. This has led to some of the world's ice melting and, in turn, the world's sea level has been rising by 1.8 mm a year over the last 50 years. If this continues, many cities in temperate regions, such as Amsterdam and New York City, will eventually be flooded.

Human interference with nature is causing the world climate to change. These changes are also affecting the land and sea. Unless pollution is controlled, these changes are likely to increase in the future.

Left *The Thames barrier is designed to stop London flooding at times of high tides and storm surges.*

Right *Polder land in the Netherlands has been reclaimed from the sea, but it will be lost if the sea level rises as a result of global warming.*

NOTE

The text on page 42 should say 'billion' where it says 'million'.

A billion is a thousand million.

Glossary

Acid A sharp, sour substance which can dissolve many metals or minerals.

Atmosphere The layer of air around the earth.

Blizzard A storm in which strong winds blow powdery snow across the land.

Celsius The Celsius scale, named after the Swedish scientist Anders Celsius, is used to measure temperature. The freezing point of water is 0° Celsius. Boiling point is 100° Celsius.

Carbon dioxide A colourless, odourless gas found in the atmosphere. Green plants need carbon dioxide to live and grow. Animals breathe in oxygen and breathe out carbon dioxide. Carbon dioxide is released when coal, oil and natural gas are burned.

Cloud A mass of tiny water droplets or ice crystals in the air. The droplets and ice crystals are so light that they do not fall to the ground.

Condensation The process by which invisible water vapour turns into a visible liquid form (as water) or a solid (as ice).

Equator A line of latitude running around the world exactly half way between the North and South Poles.

Equinoxes The two times of year when the sun is directly overhead at the equator, and when all places on earth have equal day and night – 12 hours each.

Evergreen A shrub or tree that remains in leaf throughout the year.

Fossil Evidence of ancient life found in rocks.

Front The boundary between two masses of air with different characteristics, mainly between bodies of cold and warm air.

Frosts An air temperature of 0°C or below, when water freezes. Frosts may kill plants.

Gully A deep channel worn by water, usually down a hillside.

Heath An area of uncultivated land covered by small shrubs.

Heatwave A period of unbroken hot weather, when temperatures are well above average.

Hemisphere Half a sphere.

Hurricane A storm which forms over the oceans north and south of the equator. Other names for hurricanes are tropical cyclones, typhoons and willy-willies.

Latitude Lines running around the earth parallel to the equator are called lines of latitude. They are measured in degrees between the equator (0° latitude) and the poles (90° north and south). Lines of longitude run around the world at right angles to the lines of latitude.

Ocean current A marked flow of water in the ocean. Surface currents are mainly caused by winds. Warm currents heat coastal areas. Cold currents lower temperatures.

Orbit The path followed by a heavenly body, such as the earth, around another body, such as the sun.

Pollution The fouling or harming of human, animal or plant life.

Population The total number of people who live in a place.

Scrub A type of vegetation consisting of low shrubs or stunted trees.

Solstice The time in summer or winter when the sun is overhead at its farthest northern and southern points (the Tropic of Cancer and the Tropic of Capricorn).

Subtropical region A region between the tropics and the temperate regions.

Temperature The measurement of how hot or cold something is.

Tropics The zone between the Tropic of Cancer and the Tropic of Capricorn. In the centre lies the equator.

Turbulent air Strong upward and downward currents in the atmosphere cause turbulence, creating bumpy conditions in aircraft. Turbulence occurs at ground level when winds change direction and blow in a series of gusts.

Water vapour Invisible moisture in the atmosphere, it behaves like a gas.

Further information

Books

Exploring Weather Catherall, E (Wayland Publishers Ltd, 1990)

People and Climate Hughes, J (Ginn and Company Ltd, 1993)

Endangered Forest Animals Taylor, A (Crabtree Publishing Company, 1992)

Young Geographer Investigates… Temperate Forests Jennings, T (Oxford Universtiy Press, 1991)

Videos

Climate of Concern, Shell Education Services

From Viewtech films and videos:

Acid Rain

Atmospheric Science series

Biomes series: *Temperate, Deciduous Forest*

Climate

Courses of the Seasons

Understanding Weather series

Wild Plants series: *Squirrel Woodlands*

Index - *numbers in bold show pictures*